A Beginner's Guide to
Poultry Farming in Your Backyard

Raising Chickens for Eggs and Food

Prepping and Survival Books

Dueep Jyot Singh

Mendon Cottage Books

JD-Biz Publishing

Disclaimer

The information is this book is provided for informational purposes only. It is not intended to be used and medical advice or a substitute for proper medical treatment by a qualified health care provider. The information is believed to be accurate as presented based on research by the author.

The contents have not been evaluated by the U.S. Food and Drug Administration or any other Government or Health Organization and the contents in this book are not to be used to treat cure or prevent disease.

The author or publisher is not responsible for the use or safety of any diet, procedure or treatment mentioned in this book. The author or publisher is not responsible for errors or omissions that may exist.

Warning

The Book is for informational purposes only and before taking on any diet, treatment or medical procedure, it is recommended to consult with your primary health care provider.

Our books are available at

1. Amazon.com
2. Barnes and Noble
3. Itunes
4. Kobo
5. Smashwords
6. Google Play Books

Table of Contents

Introduction

Ever since man found out that it was extremely easy to have domesticated sources of food, reared right in his yard, millenniums ago, is it a wonder that poultry especially chicken farming is one of the best methods to get easy access to a good source of food for your family?

There is absolutely no country in the world, except perhaps the Arctic regions, – where man has not reared ducks, chickens, and other poultry for table purposes down the centuries.

Apart from these being an easy source of eggs to eat for breakfast, lunch and dinner every day, you also knew that you would have a tough old rooster for dinner, when a large number of family members popped in unexpectedly, demanding sustenance.

We are going to be concentrating on chicken farming, for domestic purposes in this book. You have this dream of raising chickens in your backyard. You are interested in a continuous supply of eggs, and the occasional chicken for your pot of a Sunday. Layers are those chickens, which are normally raised for egg production. The chickens which are going to go straight into the pot are called broilers.

Since ancient times, human beings have been raising poultry for domestic purposes and also for marketing purposes.

Poultry farming has been a part of rural life in the east down the centuries. All the kitchen waste was fed to the hens. These hens came under the 21st century poultry farming term – free ranging. That meant they were allowed to scratch about in the backyard, getting their fill of insects, worms, green vegetables, organic matter, and was it a wonder that they laid delicious, nutritious, and proteinaceous eggs?

Every intelligent householder kept three or four hens depending on the size of his family, and he bought a cock from the market, when he needed chickens. Once a clutch of chickens was hatched, Cocky Locky went into the cook pot.

One of the common mistakes made by new poultry farmers is buying a large number of birds, because they are not very clear about whether they want these words for home consumption or they want to trade in the eggs and poultry meat.

Around 50 years ago, one of my father's colleagues was facing this problem. He had this huge garden and backyard. He had heard about dad rearing poultry in that garden successfully. So he also wanted to experiment in this exciting new activity which would keep his family well supplied with eggs, and fresh meat.

So the next time dad went visiting to his base on a tour, he asked dad the best way to raise birds without too much of a hassle. You are going to get these easy tips in the book.

So Mr. X bought those birds. Six months later, dad went on another tour , only to find Mr. X really discouraged. According to father, these hens should have started laying, when they were around four months old. These hens were now six months old, and Mr. X just had about 3-4 eggs per day to show for his 12 birds. They just ate their heads off and feed was so expensive.

So father sat him down and told him not to despair. How many birds did he have, 12? "How many hens?" "Six" said Mr. X., "You mean you have six roosters? What made you buy them?" Well, father had told him to buy birds, so he bought six hens and six roosters…

No wonder he was complaining about the feed. The roosters did nothing except grow fat on food, without laying eggs.

If Mr. X had been a bit practical he would have done as the ancient Egyptians did. They sold the male birds to the temples when they were three months old, for other humans to offer as sacrifices to the gods. They got their profits, they did not have these roosters taking up space in their backyards and they had plenty of hens laying eggs.

So, first rule, when you are buying birds – concentrate on hens. After all, you need eggs first. If you want to raise chickens, just buy one healthy Chanticleer for about 4 to 5 Dame Partletts. You are going to have healthy fertilized eggs with that one rooster.

The average lifespan of a chicken is about six years, but in the West, they are slaughtered, when they reach a lifespan of two years. That is because it is thought that if they have reached the limits of their egg laying potential and capacity. This of course is done for all those birds, which you have raised commercially. If you are raising them for egg laying purposes and then for table purposes, you can reap the full benefits of your chickens until they are 2 ½ years old. Then you can feast off Kentucky fried chicken grown in your own backyard.

It Is Just Chicken Feed

I am rather surprised to see that there are reports that animals and birds are fed growth hormones in farms in the West. Chickens grow really fast when they have easy access to good food. So why would they want to experiment with growth hormones? Do you want super chickens in just one week? How absurd.

Besides, these chickens are definitely not going to help their human eaters, because of the hormone content in them.

Sustainable Poultry Feed

I do not care much about extra products in the commercialized poultry feed, which is being sold in the market today. Some products are advertised with having vitamins, and other nutrients, which are going to keep your birds healthy. Remember, nothing can beat natural food. The traditional natural food, on which poultry has been flourishing in the East for millenniums is;

Equal quantities of wheat bran, Rice husk, ground corn – not finely ground in fine cornmeal – dried fish meal and rejected pieces of meat from the butcher –offal. This is the best food for having huge, delicious and totally organic eggs with a rich orange yolk.

You may want to try the above recipe, or try the husk of easily available cereals. The learned through experience episodes, which you are going to find in this book means I know of what I speak.

Apart from the food which they picked up while rooting in the garden and backyard, this extra food was enough to keep them healthy and happy. Also, father, remember to mix little pieces of stone chips in the food.

Why the stone and lime stone chips? That is because the food had to be digested in the crop before it reached the intestines. That could only be done by a grinding of the food, with the crop muscles along with the stone chips. Consider that to be natural pestle and mortar. The birds were bird witted enough not to pick up pieces of stone and grit while rooting in the yard, so he had to supplement this in the meals.

After they had eaten, he used to pick all of them, and palpitate/massage their crops with his hand. A crop which was full and in which the food moved meant that those chickens would not be "crop bound". It also meant that the digestive system was working just fine.

Crop bound Chickens

Now this is a term which many poultry farmers, especially newbies may not have come across on the Internet, very often. This normally happens when there is not much liquid matter and stone – grinding matter in the food. The food then solidifies and does not move forward. That is it, your chicken is going to die in a couple of days, because this is a very serious state.

Hand palpitation and massage of the crop, about 5 to 6 times a day after you have fed it some oil and water for lubrication, can help to alleviate this problem, but I would rather suggest prevention is better than cure.

When I was a child, father decided to teach us all about poultry farming, so that when we grew up, we knew how to domesticate birds for table purposes while getting a goodly supply of eggs every day.

These birds were of course free ranging birds. When previously he bought chicken pullets from the local government agricultural supplier, this once he told that poultry specialist to give him fertilized White Leghorn eggs. He had a brooder hen. Of course, you can use an incubator. That man just picked up some eggs, looked at their shapes and gave father 11 hens and one cock.

Now how he managed to do that, is something of which no one has managed to tell me till date, but I think it was just blind luck! We got 11 hens. But that shape idea seems to be an old wives tale.

There are people who supposedly can recognize the sex of day-old chicks, but when our chicks hatched, we tried our best to find whether they were hens or roosters. No luck! It was only when cocky started growing its red comb and tried out crowing, that we knew that he was the cock of the roost.

You do not need to have a cock in order to have hens laying eggs. They are going to start egg production, irrespective of whether there is a rooster around, when they reach around six months in a happy, healthy atmosphere.

Do not wash the eggs after you have bought them before putting them in the clutch. Bring them straight from the farm and put them under your broody or in the incubator.

Best Natural Food for Chickens

So alright, this is a tale based on experience as retold by father, long before I was born. He had just started out on his career as well as his brand-new poultry farming adventure [this was somewhere in the late 50s,] and he was winging his way through raising six White Leghorn hens all lonesome on his ownsome.

He was posted at that time in the Western part of the Indian subcontinent, which is predominantly rice eating country. So when a colleague decided to follow his example and keep hens for domestic consumption, it was "all right, this is what I have experienced, you can practice it out" as told by father.

So the colleague bought six chickens. And fed them rice, morning, afternoon and night. Along with that, these chickens were fed pieces of bread.

And they grew and they flourished, and they did not lay any eggs. Some of them died because they became crop bound.

So the next time, about six months later, when the colleague came face-to-face with father again, it was "you have given me bad advice. Egg production is nil. "

So, father quietly requested that officer to sell those chickens to father. He would practice what he preached, and prove it that his advice was good. And the chickens were bona fide good layers.

The moment the chickens came into father's hands, he immediately changed their eating patterns. Plenty of animal protein. No rice at all. Absolutely no soft floury food, which would cause them to become crop bound. Plenty of chopped greens, rice husk, and wheat bran.

Two months later, he met Mr. and Mrs. B. at a club where Mrs. B said rather archly, "so, how are *our* chickens doing?"

And father replied innocently, "Extremely well. Mrs. B. I got five eggs today, from *our* six hens. I told you they were good layers."

Instant World War III in the club itself. How could Mr. B. sell off the hens, when they were just getting ready to lay? How foolish he was. And I would not be surprised if she thought that father had pulled a fast one on simple Mr. B. She never forgave both of them ever.

It is all in the diet. Just rice and no exercise is going to give you fat birds, but not much Egg production. Also, the yolks are going to be boring and yellow. Protein diets and greens and rice husk/offal is going to give you hundred percent egg production. With rich orange delicious organic eggs.

Hatching Chickens

We had our broody hen. Father told us that white leghorns normally do not get broody – that is the reason why they are so excellent for an uninterrupted supply of eggs. But Fatty was an exception to the rule.

We learned that she was broody, when she began making a nest and sitting on pieces of stone, trying to hatch them.

So father put her in a basket, with her eggs underneath her, and food and water for easy access, next to her whenever she wanted to get up and stretch her legs. As we wanted to leave this hen undisturbed by other poultry, he placed the basket in the warm bathroom of our guestroom! "Fatty" – yes, we had named all our hens-was extremely happy there away from her friends.

Every evening he used to pick her up, and lift her off the nest. That was just in case she had been sitting on the eggs throughout the day. That would overheat them and possibly harm the chickens.

"Candling"

After that, he picked up each egg and held it up against the light bulb. Of course not very near the light bulb, which was high on the wall, but just enough to show us the shadow of the growing chicken.

In this way, he knew that the chickens were healthy in their eggs. We did not lose any chicken, but comparison between no growing chicken shadow and a clearly shown chicken shadow showed us that the eggs were healthy and none of them were addled. This is called candling.

That is because when there were no light bulbs around, poultry farmers used to look at the chicken growth with the egg held against the backdrop of candlelight.

Also, he turned the eggs by hand. The hen normally does this, when she settles down after her meal, so he knew which positions they were in, before he candled them. Then he turned them over and changed their direction.

You need to have a constant temperature, so chickens are definitely not hatched out in hot summer. Spring, autumn, especially when temperatures are mild are good times to set your broody hen on a clutch.

The temperature has to be between 98 – hundred degrees Fahrenheit, the best is, of course, 98.5°F, which is the body temperature of a hen.

The humidity level should be between 55% – 70%. Too humid an atmosphere means no chickens. Less humid an atmosphere again means no chickens.

How to Make an Incubator

Father did not go in for making an incubator, because we had just a limited number of birds. If you do not have a broody hen, you can always make an excellent incubator by going on this URL.

http://www.stormthecastle.com/how-to-make-a/how-to-make-a-homemade-egg-incubator.htm

The tips given on this URL are excellent, especially the ones on temperature control.

You may want to look for more easily made DIY incubators on the Internet, especially on earth Watch and http://www.instructables.com/id/DIY-INCUBATOR/,

where a Styrofoam box was used as a container for eggs. Excellent idea.

On the morning of the 21st day, father took us to the hen, to see how many new visitors, we had during the night. We had three in the morning, seven, When we came back from school, nine when father came back from work, and 12 by the next morning. Hundred percent success!

Hurrah, we had launched ourselves into a brand-new career, chicken farming at the age of 10 and 7 ½!

Now these baby chickens had to be raised away from the other larger poultry, until they grew up large enough to take their places among their peers.

These hens were kept in a shed. Throughout the day, they played havoc with the garden, much to the annoyance of our gardener, but he would not dare open his mouth. But all the garden pests were eaten regularly by these birds.

At this time, we had four large hens including Fatty. That is why we had to go to the agricultural farm to get fertilized eggs. This batch, of course, was going to have 11 hens and one cock.

Like you said, we were lucky and we got 11 chickens and one cock for the gardening shed of 4 m x 6 m dimensions.

There is nothing more exciting than to have baby chickens rooting about in your garden, tumbling about like little bits of fluff and cheeping after their mom. She taught them how to scratch for insect life.

She also taught them how to flatten themselves straight on the ground, the moment she gave the danger signal. Watching this is fascinating. Of course the food she taught them to collect was supplemented with food given by father in an easily accessible bowl, when they gather together in the shed at night.

And how these little chickens ate. When they were about a week old, father said that he was going to show us a very interesting experiment. We had an old Allwyn Prestcold fridge, of which the fridge lining near the door was infested with cockroaches. Father did not know how to get rid of them, because he was not going to use any pesticides near a food source.

They did not bother about the fact that the fridge was working perfectly, and the atmosphere outside the lining was freezing. So father told us to collect the chickens, switched off the fridge, and closed the kitchen door. He was going to try an experiment. And then he turned the lining and thumped on the fridge door so that the cockroaches fell out.

Those little chooks went berserk and squeaking excitedly, cockroaches! they went after them sans merci. Not a single cockroach was left however hard they tried to flee the yellow menace of these attacking relentless baby dragons.

So this is the first time I learned that little chickens love cockroaches. And they are capable of eating giant sized cockroaches in 2-3 swallows, even when they are one week old. Also, they were quite capable of jumping up and eating the cockroaches who had not fallen out of the lining. This was also the time I found out that baby chickens could jump/fly to surprisingly unexpected heights when they are in pursuit of juicy morsels.

The knowledge that this food is edible came to them instinctively. Not a single cockroach or pest dared to show its face in our house while the chicken guards roamed the rooms and kitchen.

In fact, they began to look forward to their once a week party in the kitchen, whenever father noticed more cockroach growth.

So try them as your best natural pest control experts.

Fresh Water Supply

Chickens cannot do without fresh water, so you need to give them fresh water twice a day. Why this has to be done is because they are quite capable of jumping into the water troughs and muddying the water with the dirt and dust collected from the poultry shed floor and also from the garden.

This water is normally made germ-free in the Indian subcontinent and other parts of the East with a very little bit of potassium permanganate until the water is light pink. This is because sustainable farming and farming on limited resources has been a way of life here for centuries. So poultry farmers have to make do with what they have, instead of looking for highly expensive technology.

Also, as the water container could easily be overturned by excited birds, he put a piece of brick in the container. That made it easy for the birds to dip their beaks in, and the container was steady.

Free range poultry is definitely not fenced, but they have to return to their shed for rest and laying eggs and eating their meals.

The hens are quite capable of laying their eggs on the floor, if nesting boxes are not provided.

Nesting boxes

Nesting boxes are easily provided by just turning sturdy cardboard cartons on their side, with an entrance. Make a soft nest for the hen, with straw and dried grass.

Hens normally like to roost high above the ground at night. Our shed – which happened to be the garden shed – already had a shelf about 3 feet, where the gardener could keep his implements. But they were taken over by the hens. Safety, fresh air, and natural protection was provided by the walls and the wire netting of the shed. You can use mesh walls to make coops.

Remember that the coop has to be large enough for birds to walk about freely, stretch their wings, perch, stand up, root for food and do any pleasant bird activity of which they can think, during the day.

Free Ranging Birds

We did not leave the birds free in the yard, throughout the day, very often because we were raising them in an area where natural predators were extremely common. These included Wildcats, snakes, Foxes, predatory birds on the lookout for chickens and so on.

So we just allowed them to run free in the yard, all day long, for about one hour in the afternoon or in the evening every day, depending on the weather. They could have a good time destroying the garden, rooting about for insects, before they were collected and locked up for the night.

Collecting them was rather a hectic and tiresome activity, because they just did not want to stop eating. Besides, they wanted to spend the night out!

And this brought about some rather unusual and occasionally unbelievable incidents.

We had a hen who we called Kook. And that chook was literally a kook. She had this very beguiling habit of looking at us, saying Kook? in a questioning manner, and then sitting down on the ground. We had to pick her up, and make a fuss about her.

And she would cluck happily to herself, no doubt telling the rest of the chooks- see, how well I have these human trained. I just need to say kook and flop down, and they are going to pick me up, and take me from the food and water bin to my nesting place.

Now we had a Dog called Spotty , who was trained not to run after the birds when they were in the garden. One fine afternoon, we found Spotty with a bird in his mouth. Horrified, we thought that he had killed one of our birds, because the normal instinct of a dog is to run after something which is running in front of you. And they had been running about in the farmyard, freely that afternoon.

To our amazement and relief, we found that he had brought Kook to us, live, unharmed and very pleased with herself. And so was Spotty. He had brought us a real live hen, safe and sound. She had not even been bitten because he had lifted her up, very softly, as if he was bringing us game.

As far as I can see, that bird had just taken one look at Spotty, who had just gone in that particular area for an inspection, said Kook and sat down plop on the ground. I do not know whether animals and birds can communicate with each other, but she must have told him bossily, "Do Not You Dare Hurt Me. Now Take Me to Your Leader".

And the poor perplexed Spotty did exactly that.

So I do believe that there is some wavelength going on between birds and animals on which they practice communication. Otherwise, how can we explain friendships between animals of different species like elephants and cats, horses and roosters – very common – and Spotty and Kook?

Incidentally, he was scared of all the chickens, even though he was a healthy huge dog. These white feathery things with their sharp beaks scared him awfully. But he did not mind Kook, because she used to see him around, say kook and flop down. He never picked her up again, but she could not care less. Someone else would.

And so she sat there, dusting herself in the dry dust of the yard and getting rid of all the ticks and other parasites.

Here is another surprising thing. These birds, which have been hatched and reared in your home and garden are going to be protected by other animals in your yard and farm, including your dogs and even your pet cats. They are going to be friends. Father told me this, and after seeing Spotty's protective attitude towards Kook, I believed it. He had already made up his mind that that this was a very foolish bird, and it was his job to see that nothing harmed her, when she was sitting in the garden, picking up insects around her.

That is how poultry like ducks, geese and chickens, horses, pigs and cows have lived in amity for centuries on a farm.

Dust baths and Shed Floor Covering

A bird has to have access to dry dust so that it can get rid of all the ticks. That is why your garden should have an area where dry dirt is a – plenty. All birds love dust baths. Also, you can make sure that the shed floor covering has plenty of powdery substance.

Now, why is it necessary for you to cover the floor of the shed?

Firstly, this organic covering is necessary to protect the delicate under portion of your poultry's feet. A hard ground or floor is going to hurt that area and you may find your birds suffering from a condition called bumble foot.

This floor covering is normally made up of chopped straw, organic manure- – we used cow dung – rice husk and even pieces of food and garden waste, because after all this 3 to 4 inches thick layer protected the feet of the poultry from bumble foot. Also, it gave them plenty of food material, – which included bits and pieces of food fallen from the food trough, when they were pecking in the food trough- when it was raining outside and they could not forage for themselves in the yard.

Poultry normally are very greedy birds. I remember the avidity with which all the birds used to cluck, "he is here, he is bringing us a delicious meal," whenever dad walked into the shed. In fact, I remember I and my brother laughing hilariously when one of these greedy birds would begin pecking, for the choicest pieces of fish and offal, even while father was still mixing their meals. He did it by hand, taken from the sack in which he had made a mixture of wheat bran, rice husk, corn flour, pieces of grit and stone, dried fish meal and pieces of offal., He also added crushed eggshells to this mixture, so that they could get plenty of calcium.

On that particular day, "Hungry", who could not care less that her human was still mixing the food attacked it, because she wanted to get to the meat and the fish first. And her sharp and powerful beak connected with father's hand really painfully. He immediately roared, "You. Bird."

 He was in the presence of his "babies", so he could not express himself in stronger language. The said babies were thrilled when the You Bird leapt up startled at such an unexpected show of noise from a normally silent type human, gave him a half surprised questioning "so what is the matter now" look and continued eating.

Since that day, father mixed the food outside the shed, before bringing it in. And the newly christened YouBird was the first one to jump in beak and feet first in the feeding trough.

When all your birds do that, you know that they are healthy and happy.

 Our eggs were large, delicious, with orange – yellow yolks, thanks to the rich animal and protein content, as well as greenery in the food. Out of 14 laying chickens, we had a daily average of 12 – 13 eggs or sometimes 14. This is a very excellent score.

The main problem is that our family just had four members so what did we do with 12 to 14 eggs every day? Dad could not sell those eggs or even tell our servant to sell them in the market, because he was Second in Command of that base. Besides, the question of selling eggs by Senior Ranking Officers was not considered to be worthy of consideration at that time, and era. It was just a matter of What One Could Do, and One Could Not Do.

So the neighbors benefited. They got trays of homegrown organic eggs.

Our neighbors were kind enough to send us homemade cakes, and once we got to know that this was an excellent way in which we could get delicious cakes, we solved the egg disposal problem. Okay, the eggs were two – three days old before they filled up a tray of 30, but the neighbors definitely did not mind. And besides, we got delicious cakes once or two times a week made by really good cooks.

Bumble Foot

Bumble foot is a condition when the pad of the foot gets infected. Of all our poultry, only one of them suffered from this condition, and father, being his own vet , cured her by cutting that infected area with a disinfected blade, getting rid of all the accumulated infection, washing the wound thoroughly with a potassium permanganate/Dettol solution and then bandaging the foot with a clean bandage.

She would have been cured sooner if she stayed in the box where father put her, but once the bandage was on, and she stood on her feet, assimilating the fact that her foot was not causing her pain, away she went limping into the yard. Scratching tore up the bandage, however hard and tight, that area was bound. And in the evening in she came, trailing tails of the bandage behind her. But three continuous treatments after every four days, helped her get rid of that infection.

The only problem with keeping chickens is that you so start considering them to be friends, especially when you have seen them growing from chicken-dom. You just cannot think of them as table fare. So our chickens were kept just for eggs. Also, father told me that one thing you should never do is dress a chicken which you are preparing for the table before other chickens.

Chickens are definitely not bird witted. He said that he made that mistake once and was seen by another chicken who was feeding nearby. He lifted up his head to see that chicken looking at him and looking at his erstwhile friend now being prepared for the oven.

That live chicken stopped eating and drinking, and moving about. It died of starvation and mental trauma within two weeks. And that is why Dad told us that if we wanted to make friends with our chickens, we would get disturbed when they were prepared for the table.

So the only chicken which went into the pot was Cocky. We did not mind his loss much, because he was singularly brainless, useless and also did not know when to crow. Any chicken crowing in the middle of the night, or in the middle of the day was definitely asking to be eaten.

Surprisingly enough, children are resilient creatures and we do not face any trauma when our birds died of natural causes or of accidents, like snakebite.

This of course is an example of small-scale poultry farming for home consumption. I am not going into details of large or medium scale poultry farming industries, because that is going to be rather expensive. The infrastructure in itself is going to be costly. Also, the moment you start a middle sized industry, you are going to start worrying about vet bills. You can safely keep up to 10 – 12 chickens in your backyard without bothering much about vets.

I would suggest you ask your local agricultural department about state rules and regulations, especially on the number of birds you can keep for home consumption in your yard. Do not go into the business aspect of it yet, because we are talking about easily accessible food in a sustainable environment.

Building Your Own Chicken Coop

If you want to build your own chicken coop, here is a couple of books, where you can get plenty of plans, especially if you are a DIY sort of person. We took a lot of pictures as we built our chicken coop and shared them in these books. We started with a portable chicken coop and then moved up to the bigger full chicken house so we could raise more birds.

http://www.amazon.com/s/ref=nb_sb_noss?url=search-alias%3Ddigital-text&field-keywords=john+davidson+chicken+coop+plans&rh=n%3A133140011%2C k%3Ajohn+davidson+chicken+coop+plans

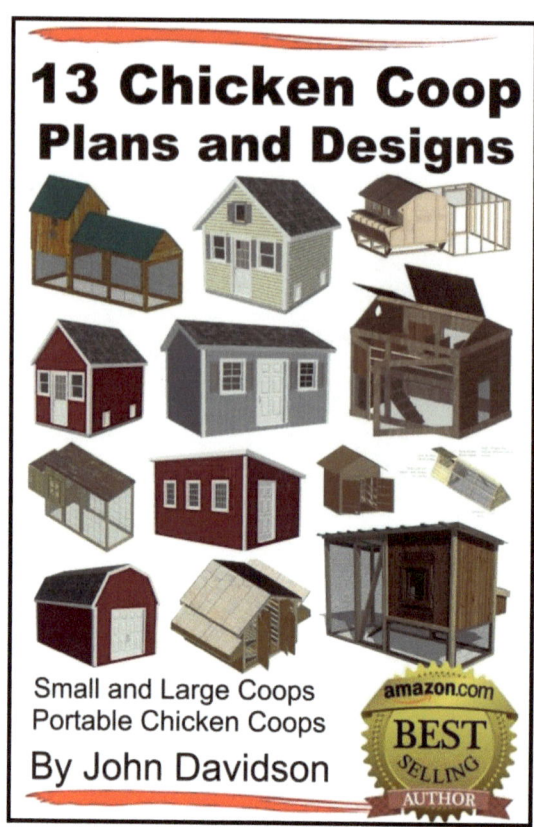

A hen house can be of any size and dimensions, depending on the number of birds you intend to raise.

Egg Production

This overcrowding means that all your chickens are now going to get their necessary amount of feed. This is going to affect egg production.

Remember that your birds need plenty of animal proteins, also greens and calcium, when they start producing eggs. That is because that calcium is going to be used to produce the eggshells. If your chickens are not fed calcium, they are going to grow weak, because the calcium from their body, which he normally should have been utilized in keeping their bones strong is going to go into the production of the eggshells. So give them protein rich and calcium rich food.

White Leghorns and Rhode Island reds are hardy species. They thrive all over the world, where there is plenty of rich organic food to eat and the weather is fine. If you can get native birds, – domestic varieties, especially jungle fowl – their eggs are even more delicious.

Once I asked a friend in Thailand, that his farm had Rhode Islands, White Leghorns and also domestic varieties. Why this eclectic mixture. Why not concentrate on just one type? He gave me a sweet smile and said, White leghorns and Rhode Islands have eggs which are larger in size. Good for market, especially for demand for organic eggs. Domestic local varieties – eggs are smaller in size, with brown shell, but excellent and delicious for feeding your family!

Raising Broilers for the Market

Broilers are chickens that you are going to raise for the market, which you are going to sell when they are about 5 to 7 weeks old. These tender and delicious birds should be about between 650 g to 850 g when fed with natural organic food. More than that, it looks like the broiler has been stuffed with potentially dangerous growth hormones or even steroids.

Unfortunately, this growth hormone and steroids trend is picking up in the East, even though the West stopped experimenting with hormones in the 1940s and steroid use in poultry feed is illegal in many states. So if you are in the East, and you are raising broilers, and talking about sustainable organic farming. Keep away from hormones and steroids. If you are in the West and believe in natural and sustainable poultry farming, I am certain you are not using these artificial growth aids like steroids and hormones.

Let me tell you something interesting here. In the East, many people believe in vegetarianism. Until 40 years ago, girls were not encouraged to eat meat. So I asked one old person about this tradition, and he said, "Ages ago, the wise men of yore found out that girls babies and girl children fed meat of male animals like cocks, bulls and rams were more aggressive in their behavior. That is why they were not encouraged to eat meat of any kind. "

Well, I can understand that sort of reasoning – animal hormones, especially when a child is reaching her teens. So they decided sensibly that boys could eat meat, because their bodies would welcome male hormones. But that did not work for girls. Estrogen does not work with testosterone, especially when it is bull, ram and rooster hormones.

Why they did not feed girls on hens or female goats – nanny goats – is something they did not bother to analyze, but then a girl child in the East was not much accepted, down the ages. So stop her from eating meat altogether.

Luckily, the 21st century female does not have to bother about such prejudice, but one may want to think of this reason why so many children are naturally aggressive. Such a simple solution. Male animal meat for boys, female animal meat for girls.

Besides, which cook is going to bother about looking at the gender of the meat being served up for dinner tonight?

Broilers can be raised in your backyard, with a floor litter in the shed made up of rice husk, wood shavings, organic fertilizer and chopped straw. The bird droppings, which fall into this litter is going to make it even more valuable as excellent garden compost. There you are, you have organic fertilizer ready at hand.

Father used to remove all this litter once a year and spread fresh litter on the floor. This organic compost was then used in our vegetable garden to give really huge and delicious vegetables.

Sustainable gardening means using all your available natural resources. So the waste material from your poultry farm is going to come very handy here.

You cannot raise broilers in small cages. Raise them in well ventilated open spaces and structures, where they have plenty of place to eat, drink, exercise and grow. These spaces should be well lit and well heated. They should also have excellent watering systems, so that you do not waste water.

For your little chickens, have small shallow troughs, in which they do not manage to drown themselves.

This is an interesting URL, which I found

http://www.the-chicken-chick.com/2012/07/the-advantages-of-poultry-nipples.html

I do not know how expensive that is, but if you are willing to experiment, why not try it to get a continuous water source, which does not waste water.

Well Ventilated Coops

When I visited the business oriented poultry farms of people I knew, I noticed that their birds were overcrowded. They had spent lots of money in making huge buildings with plenty of water and air, but the coops were overfilled with birds. Luckily the poultry farming with which I was acquainted was limited to a limited number of birds who would never suffer from ill ventilated coops and in these overcrowded and highly polluted areas.

Overcrowding is not permitted!

The pollution is due to the fact that the bird droppings mixed with ground shavings started to produce fumes of ammonia as they decomposed. This is going to affect the respiratory systems and cause problems in the development of the legs and pelvic muscles of the chickens. So your chickens are going to be more vulnerable to diseases.

Just because of overcrowding. Also, they cannot move about freely in such an atmosphere. Do you know that around 19 million birds die in the UK alone every year in these overcrowded under or overheated and ill ventilated sheds?

12 birds to one square meter means happy birds which are not going to be overcrowded. Of course, father's 4 x 6 m gardening shed was a spacious palace for 16 birds, but they would never have to worry about overcrowding. Also, this shed had plenty of natural light coming in through the wire mesh, and also there were bales of straw for when they did not want to perch on the platform. Chickens growing in such an environment may grow slower, but they are going to be healthier, and are going to produce more eggs.

Why our chickens grew so slowly was that they were free ranging birds. They were not cooped up in overcrowded pens. They had easy access to plenty of natural organic food and water. If we wanted to sell them as organic broilers, they would be ready at 12 weeks. But I remember they lived happily for four – 4 ½ years, even though the average lifespan of a white Leghorn is 5 to 6 years. And they provided us with lots and lots of eggs.

Overcrowding of chickens is also going to leave them vulnerable to chicken flu. That means your whole chicken stock is going to be wiped out through one virus, because they do not have the strong immunity needed to fight against avian flu.

Protecting chickens from Predators

Chickens in your garden are definitely going to be safe in the daytime, especially if you are on the lookout for wild cats or even hawks. Our birds were penned only at night or when the weather was miserable. This open-air access had us with really huge white Leghorns. Fatty was enormous, and much above the size and weight of an ordinary Leghorn. In fact, she would have won any fat hen competition hands down. And we had definitely no access to those harmful unnatural things called growth hormones or steroids.

So even if somebody tells you that hens do not grow more than 2.5 kgs, tell them they have known to hit 3.50 kgs for one particular Leghorn specimen. All that is due to protein and organic food. Also, fatty preferred to sit inside and eat and eat. One could almost say that she had a sedentary lifestyle. Is there a moral here somewhere?

Make sure the wire netting is strong. If you have made a chicken pen, dig a hole near the wall, and lay some netting there. That is going to prevent animals from burrowing underground and stealing away our chickens. We did not need to do that, because the shed had a solid cement floor. This naturally had to be covered with layers, so that the chickens did not hurt them when they flew off their perches in the mornings or walked on the floor during the day.

Conclusion

This book has given you lots of helpful tips to keep chickens in your backyard, especially for home purposes. This is an extremely wide-ranging subject, so apart from the tips you find here, you may want to look online for other newbies beginning to enjoy this interesting new hobby.

Keep just a limited number of birds in the initial stages.

The Truth about Growth Promoting Feed

Did you know that a large number of companies selling poultry food in the USA put antibiotics as well as drugs, which consisted of minute traces of arsenic in the feed? According to them, this drug promoted the growth of the chicken. It was only in 2000, when tests showed samples of chicken liver having enough of arsenic in it to cause neurologic problems in children or adults.

So here we have 2 ounces of chicken liver adulterated with arsenic, being fed by parents unwittingly to their children. The compound used here was and is Roxarsone. *The FDA supposedly said that the amount of arsenic found was very less in quantity than that allowed in food items.*

This makes one wonder. Does the FDA allow arsenic to be added to your food, just saying that *it is very less in quantity than what is allowed* and which is acceptable by their standards, so continue poisoning yourself.

Arsenic is a dangerous poison. You are going to be eating these chickens. FDA has deemed the element of arsenic in your food to be *acceptable* because it is less in quantity.

You think that over. FDA's statement here is quite disturbing. It shows that it does not mind trace elements of poisons placed in the food in the shape of hormones by companies producing chicken feed with growth hormones.

I would prefer you Think natural, think organic. Raise your own chickens – make your own feed.

One last point – some people advocate trimming the beak of a chicken, so that it does not peck other birds. This is almost as bad as cutting off a human's lips, because after all, what use is it, when the mouth is being used for eating purposes. Do this only if you want your chicken to suffer everlasting pain because of your mistakes – overcrowding.

All those who say that this is a painless process definitely have not seen the chickens dying, because they do not want to eat or drink anymore. How could you do that when your nerve ganglions are exposed and any sort of touch causes you unbearable pain? You would rather starve yourself, would you not.

So please do not go around trimming the beaks of your chickens.

Live healthy, live naturally. Live Long and Prosper!

Author Bio

Dueep Jyot Singh is a Management and IT Professional who managed to gather Postgraduate qualifications in Management and English and Degrees in Science, French and Education while pursuing different enjoyable career options like being an hospital administrator, IT,SEO and HRD Database Manager/ trainer, movie , radio and TV scriptwriter, theatre artiste and public speaker, lecturer in French, Marketing and Advertising, ex-Editor of Hearts On Fire (now known as Solstice) Books Missouri USA, advice columnist and cartoonist, publisher and Aviation School trainer, ex-moderator on Medico.in, banker, student councilor ,travelogue writer … among other things!

One fine morning, she decided that she had enough of killing herself by Degrees and went back to her first love -- writing. It's more enjoyable! She already has 48 published academic and 14 fiction- in- different- genre books under her belt.

When she is not designing websites or making Graphic design illustrations for clients , she is browsing through old bookshops hunting for treasures, of which she has an enviable collection – including R.L. Stevenson, O.Henry, Dornford Yates, Maurice Walsh, De Maupassant, Victor Hugo, Sapper, C.N. Williamson, "Bartimeus" and the crown of her collection- Dickens "The Old Curiosity Shop," and so on… Just call her "Renaissance Woman" - collecting herbal remedies, acting like Universal Helping Hand/Agony Aunt, or escaping to her dear mountains for a bit of exploring, collecting herbs and plants, and trekking.

Check out some of the other JD-Biz Publishing books

Gardening Series on Amazon

Download Free Books!

http://MendonCottageBooks.com

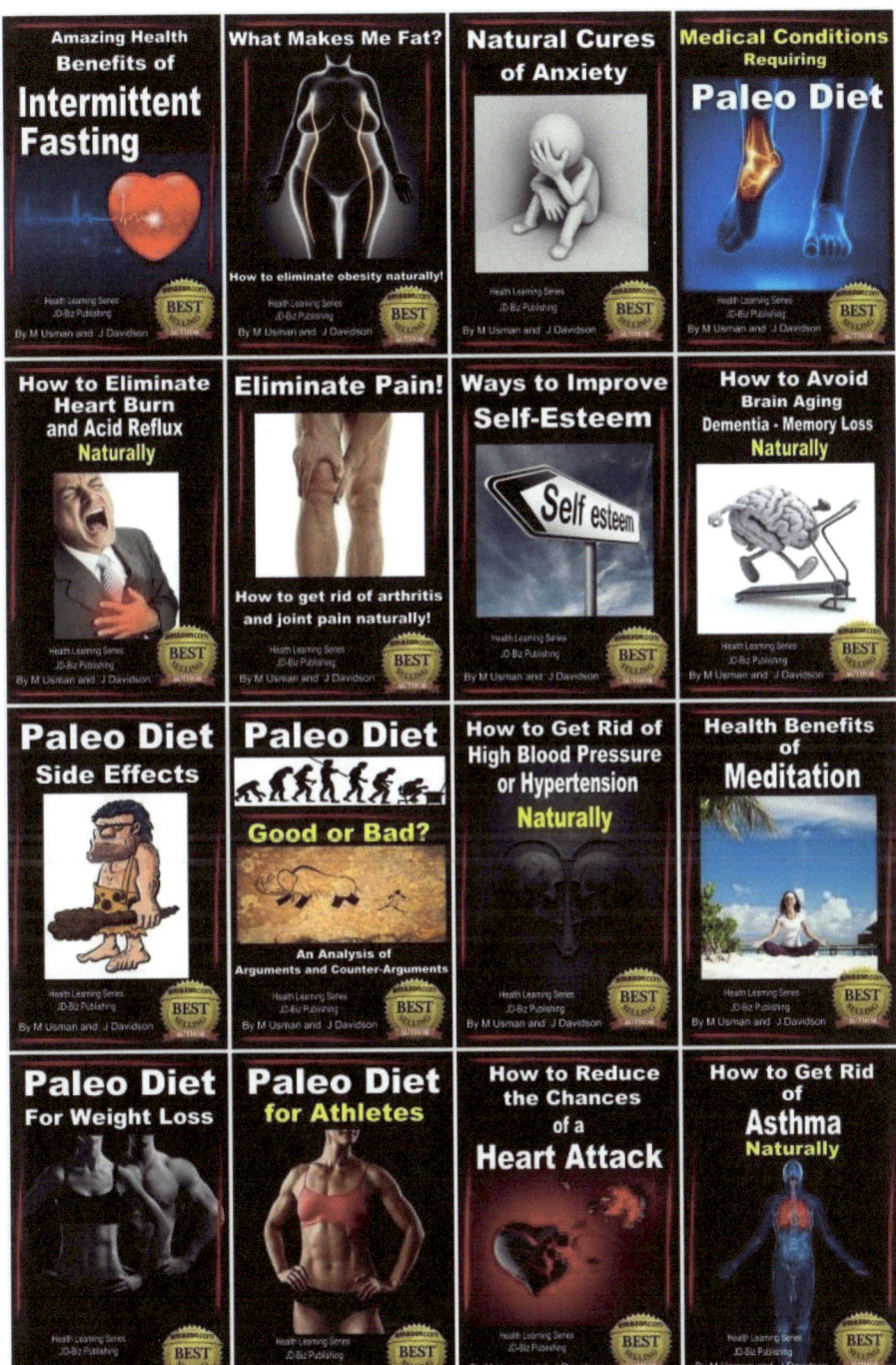

Learn To Draw Series

How to Build and Plan Books

Entrepreneur Book Series

Our books are available at

1. Amazon.com

2. Barnes and Noble

3. Itunes

4. Kobo

5. Smashwords

6. Google Play Books

Download Free Books!

http://MendonCottageBooks.com

Publisher

JD-Biz Corp

P O Box 374

Mendon, Utah 84325

http://www.jd-biz.com/

www.ingramcontent.com/pod-product-compliance
Lightning Source LLC
Chambersburg PA
CBHW040857180526
45159CB00001B/453